FLORA OF TROPICAL EAST AFRICA

MARANTACEAE

E. MILNE-REDHEAD

Perennial herbs with rhizomes, sometimes with bamboo-like shoots. Shoots simple or branched with one to many leaves. Leaves with petiole sheathing below, terete in the middle and both terete and calloused (pulvinate) just below the blade ; leaf-blade usually more or less asymmetric with one side (either left or right) more curved than the other and the acumen not in line with the midrib ; some plants with only one type of leaf (the more curved side either right or left) (homotropic), some with both types (antitropic) ; nerves parallel and very numerous. Inflorescence simple or branched with each cymose partial inflorescence in the axil of a primary bract, each successive branch of the cyme (a two-flowered cymule) enclosed in an unkeeled or 2–3-keeled prophyll, and an unkeeled mesophyll opposite the prophyll, or with mespohyll or both mesophyll and prophyll absent ; primary bracts arranged in simple spikes with from 4 to many nodes. Cymule pedunculate or sessile with the two flowers pedicellate or seesile, side by side or at different levels, with or without fleshy bracteoles. Sepals free, equal. Corolla forming a tube below with 3 petaloid lobes. Staminodes (with stamen) forming a tube below, which is fused to the corolla tube ; outer staminodes 2 or 1, petaloid or subulate, sometimes unequal ; inner staminodes 3, unequal, one bearing the anther and another hooded, with or without a spur-like appendage. Style and stigma at first held erect by the hood of one of the staminodes, later suddenly bending downwards. Ovary inferior, 1- or 3-locular with one ovule in each loculus. Fruit dehiscent or indehiscent, with 1–3 seeds. Seed with or without a basal aril ; endosperm abundant.

Plants with bamboo-like shoots ; leaves arranged distichously on leafy shoots ; inflorescence branches with up to 20 nodes ; each two-flowered cymule with 2 fleshy bracteoles ; ovary papillose ; capsule muricate ; seed arillate . 1. **Trachyphrynium**

Plants without bamboo-like shoots ; leaves not arranged distichously on leafy shoots :

Stems simple :

Leaf-blades scarcely asymmetric ; each cymule with one or two fleshy bracteoles ; fruit with fleshy pulp inside :

Nodes of inflorescence each with one cymule ; each bract enclosing the part of the inflorescence above it, deciduous at anthesis ; each cymule with one fleshy bracteole ; fruit with well-marked sutures ; seed arillate . . . . 2. **Megaphrynium**

Nodes of inflorescence each with 2–4 cymules ; bracts not enclosing the part of the inflorescence above it, persistent ; each cymule with two fleshy bracteoles ; fruit without obvious sutures ; seed without an aril . . . . 3. **Sarcophrynium**

Leaf-blade strongly asymmetric ; cymules without fleshy bracteoles ; fruit tardily dehiscent, without fleshy pulp inside ; seed arillate . 4. **Marantochloa**

Stems branched ; leaf-blades strongly asymmetric ; cymules without fleshy bracteoles; seed arillate:

Bracts acute ; prophyll not two-keeled ; flowers at different levels ; fruit tardily dehiscent . 4. **Marantochloa**

Bracts obtuse ; prophyll two-keeled ; flowers ± side by side ; fruit indehiscent . . . 5. **Ataenidia**

The above 5 genera all have a 3-locular ovary. The genus *Thalia* L., which is a plant of swamps in open country and which has not yet been found in the area covered by this Flora, has a unilocular ovary. *Thalia* is known from A.-E. Sudan, Belgian Congo and Northern Rhodesia, and may be expected to occur in Uganda and western Tanganyika.

1. TRACHYPHRYNIUM

Benth. in G.P. 3 : 651 (1883)

Hybophrynium K. Schum. in E.J. 15 : 428 (1892) ; E.P. IV. 48 : 41 (1902)

Woody herbs with bamboo-like shoots. Shoots erect or supported by other plants, simple below and clothed with sheathing coriaceous cataphylls, branched and leafy above. Leaves petiolate, distichous ; petioles sheathing below, jointed shortly above the top of the sheath, terete and calloused above the joint, joining the under surface of the midrib without interruption ; leaf-blades asymmetric, antitropic. Inflorescence terminal, simple or sometimes branched at the base, spike-like with a jointed rhachis and a two-flowered cymule at each distichous node, subtended by a sheathing bract enclosing the cymule and the part of the inflorescence above it ; bract deciduous at the time of flowering ; prophylls and mesophylls absent. Cymule shortly pedunculate, with the flowers side by side, each subtended by a small fleshy bracteole. Outer staminodes 2, petaloid ; the hooded staminode with a spur-like appendage. Ovary 3-locular, completely clothed by contiguous papillæ. Capsule dehiscent, muricate, 1–3-seeded. Seed smooth, with a basal aril.

T. braunianum (*K. Schum.*) *Baker* in F.T.A. 7 : 319 (1898) ; Milne-Redh. in K.B. 1950 : 158 (1950) ; B.S.B.B. 83 : 7, 11 (1950). Type : A.-E. Sudan, Equatoria Province, Atasilli Brook, *Schweinfurth* 3061 (B, lecto., K, iso.-lecto. !)

Rhizome creeping, much branched. Bamboo-like shoots 2–4·5 m. or occasionally up to 7·5 m. high, forming thickets. Cataphylls persistent, up to 21 cm. long. Leaves very variable in size ; blades elliptic, slightly asymmetrical, shortly acuminate, rounded or cordate at the base, held ± in two ranks with the more curved margin nearer to the stem, ranging from 5·5 × 2·0 cm. and 12·5 × 4·0 cm. to 20 × 7·5 cm. and 19 × 10 cm. ; petiole above the joint 0·5–1·7 cm. long. Inflorescence up to 20 cm. long with up to 43 nodes, usually shorter with fewer nodes ; internodes 4–7 cm. long, shortly pubescent ; bracts 2·0–2·8 cm. long ; peduncle 2–4 mm. long ; fleshy bracteoles about 2 mm. long. Flowers white, sometimes tinged pink or purple ; expanded flower about 2 cm. in diameter. Capsule lobed if more than 1-seeded, obovoid or the lobes ± obovoid, about 1·5 cm. long, strongly muricate, orange (see Fig. 1/1) ; seed obovoid, about 7 mm. diameter, glossy black or somewhat brownish, with a brownish-white aril. Fig. 1/1 and 1A.

UGANDA. Masaka District : Sese Islands, Bugala, 3 June 1932 (fl.), *Thomas* 27 ! ; Mengo District : Entebbe, Kitubulu, June 1946 (fl. & fr.), *Eggeling* 5707 ! Entebbe Mar. 1935 (fl. & fr.), *Chandler* 1172 !
DISTR. **U**2, 4 ; Sierra Leone to Belgian Congo and A.-E. Sudan
HAB. Wet places in rain-forest ; 1010–1170 m.

SYN. *Hybophrynium braunianum* K. Schum. in E.J. 15 : 428, fig. A–E (1892) ; E.P. IV. 48 : 41 (1902) ; Hutch. in F.W.T.A. 2 : 336 (1936)

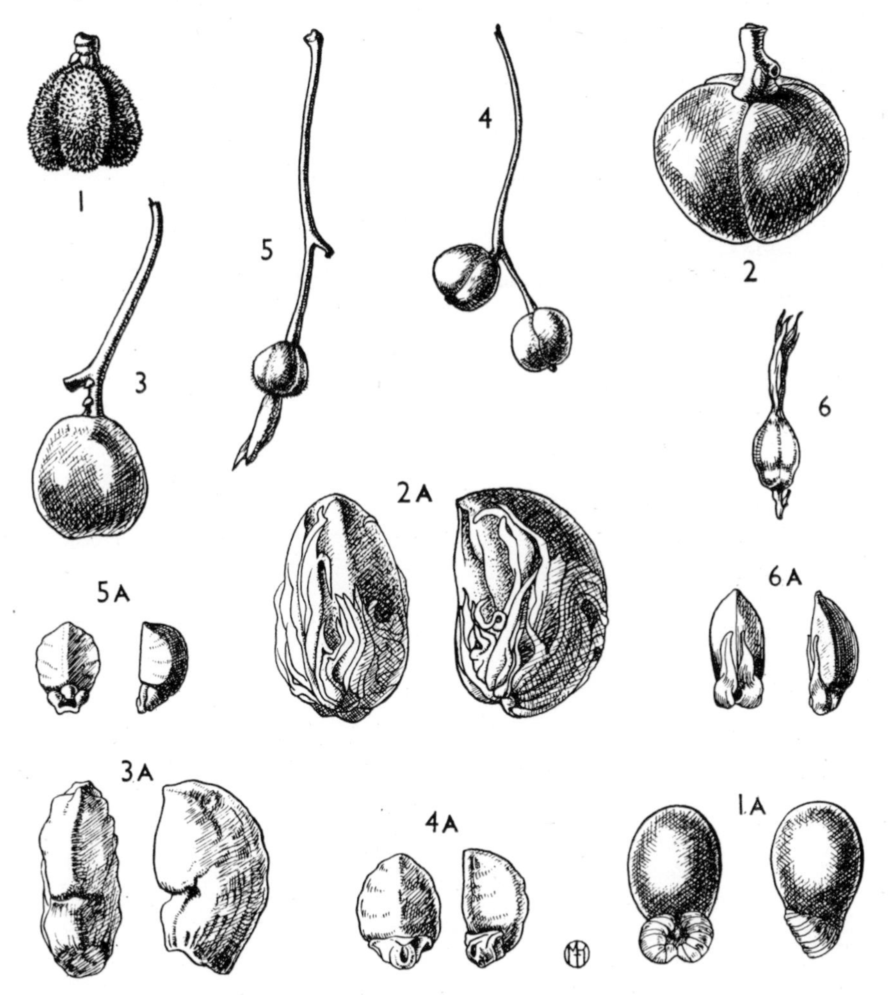

FIG. 1. FRUITS (× 1) AND SEEDS (× 2) OF *MARANTACEAE*—1, 1A, Trachyphrynium braunianum ; 2, 2A, Megaphrynium macrostachyum ; 3, 3A, Sarcophrynium schweinfurthianum ; 4, 4A, Marantochloa leucantha ; 5, 5A, Marantochloa purpurea ; 6, 6A, Ataenidia conferta.

2. MEGAPHRYNIUM

Milne-Redh. in K.B. 1952 : 169 (1952)

Sarcophrynium K. Schum. in E.P. IV. 48 : 35 (1902), *pro parte*

Herbs with rhizome, glabrous or pubescent. Stems simple bearing an inflorescence and a single subtending leaf. Other leaves arising direct from the rhizome. Petiole sometimes very long (up to 5 m.) ; leaf-blade elliptic, scarcely asymmetric. Inflorescence terminal, branched, the branches spike-like, jointed with many internodes and a two-flowered cymule at each node, subtended by a sheathing bract enclosing the cymule and the part of the inflorescence above it ; bract deciduous at the time of flowering ; each cymule backed by a two-keeled prophyll ; mesophylls absent. Cymule pedunculate with the flowers side by side with one fleshy bracteole between them. Outer staminodes 2 or 1, narrowly lanceolate, linear or subulate ; the hooded staminode with a spur-like appendage. Ovary 3-locular, glabrous or pubescent. Fruit fleshy, indehiscent but with 3 conspicuous sutures. Seed with an aril.

M. macrostachyum (*Benth.*) *Milne-Redh.* in K.B. 1952 : 170 (1952). Type : British Cameroons : Ambas Bay, *Mann* 1335 (K, holo. !)

Glabrous or pubescent herb ; rhizome up to 6 m. long bearing singly numerous leaves. Stem up to 2·5 m. long. Leaves with petiole up to 5 m. long but often less ; calloused part 7–15 cm. long ; blade ovate-elliptic, ± symmetric, normally 6 dm., occasionally up to 9 dm. long, acute or shortly acuminate, ± rounded at the base. Inflorescence arising 10–33 cm. below the calloused part of the petiole, up to 25 cm. long ; internodes up to 20, 4–8 mm. long ; bracts 17–23 mm. long, white ; peduncle 6–10 mm. long ; fleshy bracteole about 2·5 mm. long. Flowers white, tinged yellow or orange, when expanded about 2 cm. in diameter. Fruit depressed globose, about 2·5 cm. in diameter, smooth, bright scarlet, with white pulp inside, normally 3-seeded ; seed as in Fig. 1/2A, p. 3, about 15 mm. long, chocolate-purple, bluish-violet or black, clothed in a deeply laciniate whitish aril. Fig. 1/2 and 2A, p. 3.

UGANDA. Toro District : Kilemba, 20 Febr. 1939 (fr.), *Thomas* 2776 ! ; Mengo District : Entebbe Botanic Garden, probably native, Apr. 1948 (fl.), *Eggeling* 5770 ! & 15 July 1948 (fr.), *Eggeling* 5770A !

DISTR. **U**2, 4 ; Sierra Leone to Belgian Congo and A.-E. Sudan

HAB. Wet places in rain-forest ; 1130–1500 m.

SYN. *Phrynium macrostachyum* Benth. in G.P. 3 : 653 (1883)
P. benthami Baker in F.T.A. 7 : 323 (1889), *nom. illegit.* Type as *M. macrostachyum* (Benth.) K. Schum.
Sacrophrynium macrostachyum (Benth.) K. Schum. in E.P. IV. 48 : 37 (1902) ; Hutch. in F.W.T.A. 2 : 336 (1936) ; B.S.B.B. 83 : 28 (1950)
Sarcophrynium arnoldianum De Wild. in Ann. Mus. Congo., Bot. Sér. V, 1 : 107 (1904). Type : Belgian Congo, Kasai, Lusanga, *Gentil* 85 (BR, lecto.)

3. SARCOPHRYNIUM

K. Schum. in E.P. IV. 48 : 35 (1902), *pro parte*

[*Phrynium* sensu Baker in F.T.A. 7 : 321 (1898), *pro parte, non Willd.*]

Glabrous or pubescent herbs. Stem simple, bearing inflorescence and a single subtending leaf. Other leaves arising direct from the rhizome or at the base of the stem. Leaf-blade elliptic or oblong-elliptic, scarcely asymmetric. Inflorescence terminal, branched, the branches with few to many nodes, and with up to 4 two-flowered cymules at each node, subtended by a sheathing bract which does not enclose the part of the inflorescence above it, not deciduous at time of flowering ; each cymule backed by a two-nerved prophyll ; mesophylls absent. Cymule pedunculate with the flowers at different levels, each subtended by a fleshy bracteole. Outer staminodes 2, petaloid ; the hooded staminode with a spur-like appendage. Ovary 3-locular, glabrous. Fruit fleshy, indehiscent, without obvious sutures. Seed without an aril.

S. schweinfurthianum (*O. Ktze.*) *Milne-Redh.* in B.S.B.B. 83 : 30 (1950). Type : A.-E. Sudan, Equatoria Province, Dyagbe River, *Schweinfurth* 3103 (K, holo. !)

Herb with rhizome up to 2 m. long bearing leaves in tufts surrounding the leaf subtending the inflorescence. Stems ± 40 cm. long. Leaves with a petiole up to 1·5 m. long, usually less ; calloused part 3–11 cm. long ; blade elliptic, slightly asymmetric, up to 5 dm. long, shortly acuminate, ± rounded at the base. Inflorescence arising 15–60 cm. below the calloused part of the petiole of the subtending leaf, up to 20 cm. long, branched ; internodes 1·2–2·0 cm. long ; cymules fascicled at each node ; bracts 2·2–3·0 cm. long, persistent, pinkish ; peduncle 2·2–3·0 cm. long, enlarging somewhat in fruit ; fleshy bracteoles about 1 mm. long. Flowers white with red patch in throat, when expanded about 12 mm. in diameter. Fruit globose pyriform, about 15 mm. in diameter, 1–3-seeded, smooth, bright scarlet with a white sticky pulp inside ; seed as in Fig. 1/3A, p. 3, about 13 mm. long. Fig. 1/3 and 3A, p. 3.

Uganda. Masaka District : Sese Islands, Bugala, 25 Febr. 1945 (fl. & fr.), *Greenway & Thomas* 7177 ! ; Mengo District : Entebbe, Kitubulu, Febr. 1948 (fl. & fr.), *Eggeling* 5750 ! & Entebbe, Oct. 1947 (fl. & fr.), *Eggeling* 5721 !

Distr. **U**2, 4 ; Belgian Congo and A.-E. Sudan

Hab. Swamp-forest on sand ; 1120–1140 m.

Syn. *Arundastrum schweinfurthianum* O. Ktze., Rev. Gen. 2 : 684 (1891), *pro max. parte, quoad fr.*
Phyllodes baccatum K. Schum. in E.J. 15 : 442 (1892). Type : Belgian Congo, Kasai, Mukenge, *Pogge* 1439 (B, holo. †)
Sarcophrynium baccatum (K. Schum.) K. Schum. in E.P. IV. 48 : 39 (1902)

4. MARANTOCHLOA

[Brongn. ex] Gris in B.S.B.F. 7 : 321 (1860)

Clinogyne Benth. in G.P. 3 : 651 (1883)

Glabrous or pubescent herbs. Stems erect or spreading, branched or rarely simple, bearing many leaves or rarely only one leaf, glabrous or pubescent. Leaf-blade strongly asymmetric, homotropic or rarely antitropic, the rounded side with a more deeply coloured marginal zone beneath (i.e. the part exposed in bud). Inflorescence lax or congested, each branch with about four nodes, with one or two fascicled cymules at each node ; each cymule backed by a prophyll ; mesophylls present ; cymule pedunculate or subsessile, with the pedicellate flowers arranged at different levels ; fleshy bracteoles absent. Outer staminodes 2, petaloid ; the hooded staminode with a spur-like appendage. Ovary 3-locular, often pilose. Capsule tardily dehiscent, smooth, more or less pubescent, not fleshy within. Seed shaped like a third segment of a sphere, with a small basal aril.

Inflorescence lax ; peduncle of cymule over 2 cm. long :
 Terete middle part of petiole up to 1 cm. long ; flowers about 0·7 cm. long ; perianth deciduous in frut . . . . 1. *M. leucantha*

Terete middle part of petiole up to 38 cm. long ; flowers about 1·8 cm. long ; perianth persistent in fruit . . . . 2. *M. purpurea*
Inflorescence congested ; peduncle of cymule subsessile or under 1·0 cm. long . . . . 3. *M. mannii*

1. **M. leucantha** (*K. Schum.*) *Milne-Redh.* in B.S.B.B. 83 : 19 (1950). Type : British Cameroons, Barombi [Kumba], *Preuss* 495 (B, holo., K, iso. !)

Stems up to 4·0 m. high but often much less, branched. Leaves homotropic ; sheathing part of petiole up to 24 cm. long ; terete part below the calloused part less than 1 cm. long ; calloused part up to 2·0 cm. long ; leaf-blade strongly asymmetric, the acumen normally being to the right of the midrib as seen from above, ± ovate, up to 25 × 14 cm., often much smaller, sometimes purplish below. Inflorescence lax, up to 30 cm. long, branched ; rhachis and bracts green ; primary bract 2·5–3·5 cm. long, spreading with the pedunculate cymules in anthesis ; flowers creamy or greenish-white, about 0·7 cm. long ; ovary shortly pilose. Capsule subglobose, about 9 mm. in diameter, glabrescent, glossy, bright red or creamy white, the withered perianth deciduous ; seed as in Fig. 1/4A, p. 3, brownish or greyish with an aril. Fig. 1/4 and 4A, p. 3.

UGANDA. Toro District : Bwamba, Bulanga, 30 Sept. 1932 (fr.), *Thomas* 754 ! ; Kigezi District : Malambagambo Forest, Febr. 1950 (fl.), *Purseglove* 3294 ! ; Mengo District : Entebbe, Sept. 1947 (fr.), *Eggeling* 5715 !
TANGANYIKA. Lushoto District : S. of Amani, between Kwamkoro and Ngua, 13 Dec, 1936 (fl. & fr.), *Greenway* 4816 ! & Amani, near R. Kwamkuyu, 13 May 1950 (fr.), *Verdcourt* 193 ; Morogoro District : Nguru Mts., Mhonda, Apr. 1892, *Sacleux* 1832 !
DISTR. **U**2, 4, **T**1, 3, 6 ; Sierra Leone to Angola, Belgian Congo and A.-E. Sudan
HAB. Clearings and secondary growth in rain-forest ; 750–1200 m.

SYN. *Donax leucantha* K. Schum. in E.J. 15 : 436 (1892) ; Baker in F.T.A. 7 : 317 (1898)
Clinogyne ugandensis K. Schum. in P.O.A. C : 150 (1895) ; Baker in F.T.A. 7 : 318 (1898). Type : Uganda, Kigezi or Ankole District, Maryonyo, *Stuhlmann* 1390 (B, holo.)
Clinogyne leucantha (K. Schum.) K. Schum. in E.P. IV. 48 : 66 (1902)
[*Marantochloa flexuosa* sensu Hutch. in F.W.T.A. 2 : 338 (1936), *pro parte, non Phrynium flexuosum* Benth.]

2. **M. purpurea** (*Ridl.*) *Milne-Redh.* in B.S.B.B. 83 : 21 (1950). Type : Angola, Golungo Alto, *Welwitsch* 6440 (BM, holo. !)

Stems up to 3·0 m. high but often much less, branched. Leaves homotropic ; sheathing part of petiole up to 42 cm. long ; terete part below the calloused part up to 38 cm. long ; calloused part up to 4·0 cm. long ; leaf-blade strongly asymmetric, the acumen normally being to the right of the midrib as seen from above, ± ovate, up to 43 × 18 cm., often much smaller, sometimes pruinose or purplish below. Inflorescence lax, up to 45 cm. long, branched ; rhachis and bracts pink ; primary bract 2·5–4 cm. long, spreading, together with 2 two-flowered pedunculate cymules which it envelopes. Flowers pale pink or deep purple with two inner lobes of staminode bright yellow, about 1·8 cm. long ; ovary shortly pilose. Capsule subglobose, about 8 mm. in diameter, pilose, bright red, the withered perianth persisting ; seed as in Fig. 1/5A, brown, smooth with whitish aril. Fig. 1/5 and 5A, p. 3, and Fig. 2.

UGANDA. Masaka District : Sese Islands, Bufumira, July 1945 (fl. & fr.), *Purseglove* 1717 ! ; Mengo District : Entebbe, Sept. 1947 (fl. & fr.), *Eggeling* 5716 ! & Kyewaga Forest, 1 Sept. 1949 (fl. & fr.), *Dawkins* 346 !

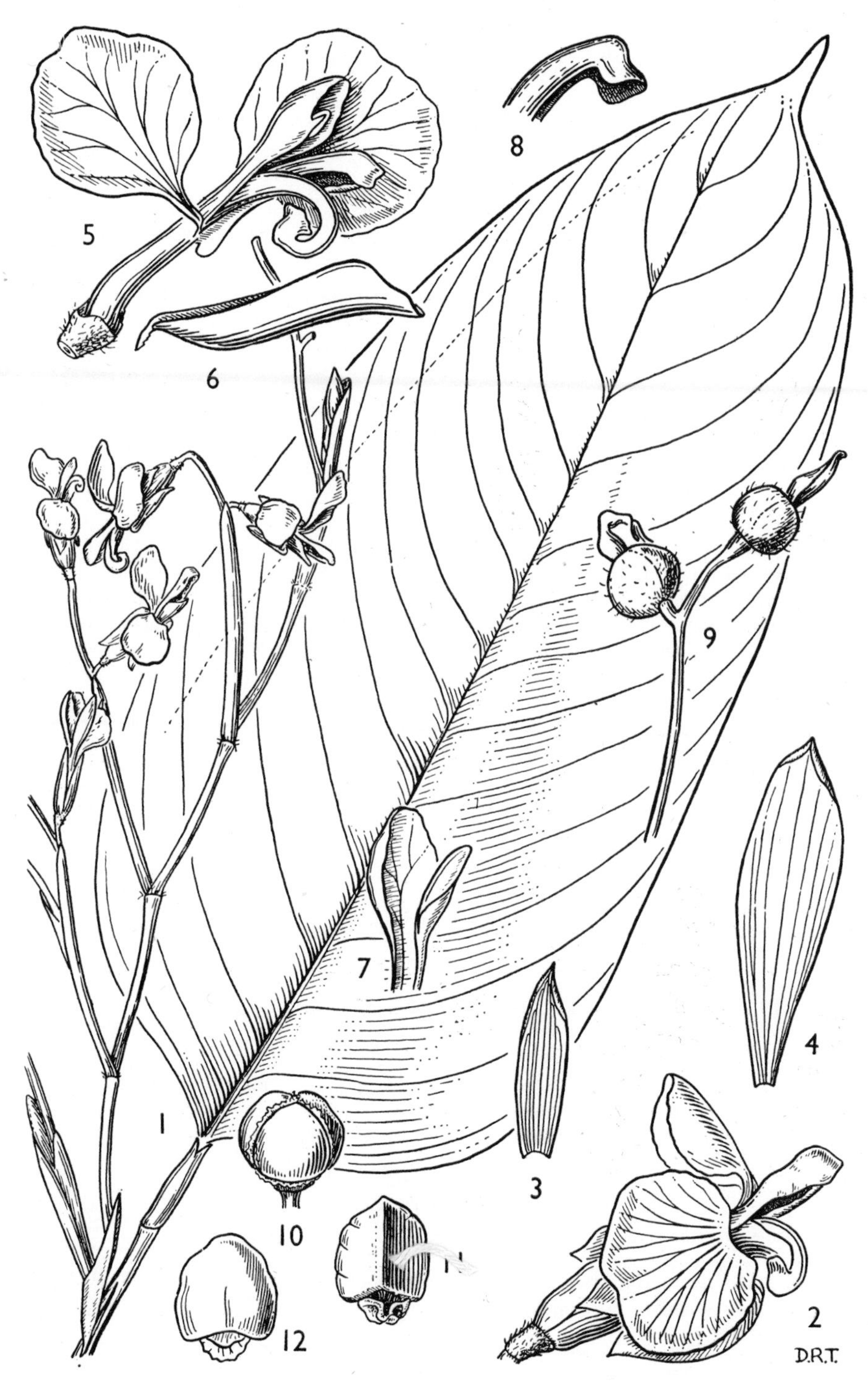

FIG. 2. *MARANTOCHLOA PURPUREA* (mainly from Thomas 2774 from U2)—1, part of inflorescence and upper leaf, × 1 ; 2, flower with style reflexed, × 3 ; 3, sepal, × 3 ; 4, corolla lobe, × 3 ; 5, flower with perianth and one inner staminode removed to show style, × 3 ; 6, the inner staminode removed from 5, × 3 ; 7, stamen, × 4 ; 8, upper part of style and stigma, × 4 ; 9, fruiting cymule, × 1 ; 10, fruit with valves removed to show seeds, × 2 ; 11, 12, dorsal and ventral view of seed, × 3.

TANGANYIKA. Bukoba District : Kikuru Forest, 16 Sept. 1934 (fl.), *Gillman* 158 ! & Munene Forest, 9 Febr. 1924 (fl.), *Grant* 124 !
DISTR. **U**2, 4, **T**1 ; Sierra Leone to Angola, Belgian Congo and A.-E. Sudan
HAB. Clearings in rain-forest ; 1110–1500 m.

SYN. *Clinogyne purpurea* Ridl. in J.B. 25 : 132 (1887)
Donax purpurea (Ridl.) K. Schum. in E.J. 15 : 440 (1892) ; Baker in F.T.A. 7 : 318 (1898) ; Rendle in Cat. Welw. Afr. Pl. 2 : 23 (1900)
Clinogyne baumannii K. Schum. in E.P. IV. 48 : 63 (1902). Type : Sierra Leone, Nienia, *Scott Elliot* 4902 (B, lecto., K, iso.-lecto. !)
[*Marantochloa flexuosa* sensu Hutch. in F.W.T.A. 2 : 338 (1936), *pro parte, non Phrynium flexuosum* Benth.]

VARIATION. The flower-colour of all specimens seen from Uganda and Tanganyika is pale pink, whilst specimens from West Africa and Angola have deep purple flowers. It seems that flower-colour in *M. purpurea* is correlated with geographical distribution, as is fruit-colour in *M. leucantha*.

3. **M. mannii** (Benth.) *Milne-Redh.* in K.B. 1952 : 167 (1952). Type : Fernando Po, *Mann* 1173 (K, holo. !)

Stems up to 2·0 m., usually about 1·5 m. high, branched. Leaves homotropic ; sheathing part of petiole up to 24 cm. long, terete part below the calloused part up to 15 cm. long ; calloused part up to 3·5 cm. long ; leaf-blade strongly asymmetric, the acumen normally being to the right of the midrib as seen from above, ± ovate, up to 46 × 20 cm., often much smaller, often glaucous or purplish below. Inflorescence congested, branched, up to 9 cm. long ; rhachis and bracts pink or bright red ; primary bracts 3–3·8 cm. long, scarcely spreading, concealing the cymule ; cymule subsessile or with a peduncle up to 6 mm. long. Flowers white or pale pink, about 18 mm. long ; ovary shortly pilose. Capsule and seed unknown.

UGANDA. Masaka District : Sese Islands, Bukasa, 27 Febr. 1945 (fl.), *Greenway & Thomas* 7204 ! ; Mengo District : Nakiza Forest, 24 Jan. 1951 (fl.), *Dawkins* 698 ! & Entebbe, Dec. 1947 (fl.), *Eggeling* 5724 !
TANGANYIKA. Bukoba District : Kikuru Forest, 16 Sept. 1934 (fl.), *Gillman* 158A ! & Munene Forest, 1925, *Wigg* 272 !
DISTR. **U**2, 4, **T**1 ; Gold Coast to Belgian Congo
HAB. Rain-forest ; 1100–1200 m.

SYN. *Calathea mannii* Benth. in G.P. 3 : 653 (1883) ; Baker in F.T.A. 7 : 327 (1898)
Phrynium hensii Baker in F.T.A. 7 : 323 (1898). Type : Belgian Congo : Bangala, *Hens* 140 (K, holo. !)
Phrynium mannii (Benth.) K. Schum. in E.P. IV. 48 : 56 (1902) ; Hutch. in F.W.T.A. 2 : 337 (1936)
Clinogyne hensii (Baker) K. Schum. in E.P. IV. 48 : 62 (1902)
Marantochloa hensii (Baker) Pellegr. in Mém. Soc. Linn. Norm., Nouv. Sér., Bot., 1 : 45 (1938) ; B.S.B.B. 83 : 22 (1950)

NOTE. Mr. H. C. Dawkins states that *M. mannii* is established in the Entebbe Botanic Garden, where it grows in a luxuriant stand and flowers freely. He has been unable to find fruit, however, and suggests that the species may require wetter conditions.

5. **ATAENIDIA**

Gagnep. in B.S.B.F. 55 : XLI (1908)

[*Calathea* sensu Baker in F.T.A. 7 : 326 (1898), *pro parte, non G. F. Mey.*]
[*Phrynium* sensu K. Schum. in E.P. IV. 48 : 52 (1902), *pro parte, non Willd.*]

Herbs with rhizome. Stems erect or spreading, branched, bearing one inflorescence and several leaves. Petiole sometimes long (up to 9 dm.), the part directly below the leaf-blade calloused ; blade elliptic, asymmetric. Inflorescence terminal, much branched, very condensed, the branches with few nodes and with up to 4 two-flowered cymules at each node, subtended by a wide sheathing bract which does not enclose the part of the inflorescence above it, persistent ; each cymule backed by a two-nerved prophyll ; mesophylls present. Cymule subsessile with the flowers ± side by side but opening at different times ; fleshy bracteoles absent. Outer staminodes 2, petaloid ; spur-like appendage to hooded staminode absent. Ovary 3-locular, pubescent. Fruit indehiscent, 1–3-seeded. Seed with an aril.

Ataenidia conferta (*Benth.*) *Milne-Redh.* in K.B. 1952 : 168 (1952). Type : British Cameroons, Cameroon Mt., *Mann* 2444 (K, holo. !)

Herb with short rhizome and tufted habit. Stems up to 6 dm. long. Leaves with a petiole up to 9 dm. long, but often less ; calloused part of petiole up to 5 cm. long ; blade elliptic, asymmetric, antitropic, up to 48 cm. long, acuminate, ± rounded at the base, often tinged reddish below. Inflorescence one per stem, appearing to be axillary, surrounded by up to 6 leaves, much branched, the branches congested with short internodes, subtended by broadly elliptic obtuse or acute strongly imbricate bracts ; cymules about 4 at each node, enclosed in a similar broadly elliptic bract about 2·5 cm. long ; bracts various shades of red, held vertically, horizontally or more or less inverted, depending on their position in the inflorescence ; cymules sessile ; ovary pubescent. Flowers pale pink or pale purple, sometimes white, about 2 cm. long. Fruit ellipsoid, obscurely lobed if more than one-seeded, about 9 mm. long and 6 mm. in diameter, not fleshy, with the withered perianth persisting ; seed as in Fig. 1/6A, p. 3, brown with a whitish aril. Fig. 1/6 and 6A, p. 3.

Uganda. Masaka District : Sese Islands, Towa Forest, 27 July 1939 (fl.), *Thomas* 3021 ! ; Toro District : Nabulongwe Forest, 27 km. W. of Fort Portal, 19 Dec. 1949 (fl. & fr.), *Dawkins* 478 ! ; Mengo District : Entebbe, Oct. 1947 (fl. & fr.), *Eggeling* 5717 !

Distr. **U**2, 4 ; Gold Coast to Belgian Congo and Angola

Hab. Wet places in rain-forest ; 840–1170 m.

Syn. *Calathea conferta* Benth. in G.P. 3 : 653 (1883) ; Baker in F.T.A. 7 : 327 (1898)
Phrynium confertum (Benth.) K. Schum. in E.P. IV. 48 : 56 (1902) ; Hutch. in F.W.T.A. 2 : 337 (1936) ; B.S.B.B. 83 : 24 (1950)
Ataenidia gabonensis Gagnep. in B.S.B.F. 55 : XLI (1908). Type : Gabon, Tchibanga, *Le Testu* 1154 (P, lecto., K, iso.-lecto. !)

INDEX TO MARANTACEAE